Bibliographic information published by the German National Library:

The German National Library lists this publication in the National Bibliography; detailed bibliographic data are available on the Internet at http://dnb.dnb.de .

Imprint:

Print and binding: Books on Demand GmbH, Norderstedt Germany
ISBN: 9783656987208

This book at GRIN:

http://www.grin.com/en/e-book/337001/revisiting-constraints-towards-cautious-reducing-emissions-from-deforestation

Omar Pidani

Revisiting Constraints: Towards Cautious Reducing Emissions from Deforestation and Forest Degradation (REDD) Implementation in Indonesia

GRIN Publishing

Revisiting Constraints: Towards Cautious Reducing Emissions from Deforestation and Forest Degradation (REDD) Implementation in Indonesia[1]

Abstract

As one amongst few countries with largest forest coverage, Indonesia puts a lot of hopes of benefitting from REDD Implementation, a program strongly believed as incentive for forest protection in developing countries. Yet, like many programs in the past offering similar incentives, some doubt that REDD will be smoothly implemented. This paper tries to revisit some of major obstacles put forward by scholars and practitioners. These include the issue of leakage and land tenure insecurity, as well as the issue of carbon pricing which are crucial in determining whether or not incentive offered by REDD is more attractive than incentives to deforest. Discussion about factors that contribute to leakage and land tenure insecurity is then followed by simple calculation to predict whether price of carbon credit per hectare offered at REDD Project in Ulumasen Forest is much more interesting than revenue obtained from logging that one hectare. This calculation indicated that incentive to forest logging surpassed incentive to conserve as a result of carbon credit. It is recommended that careful and transparent feasibility study prior to REDD implementation to avoid its inevitable marginalizing impacts particularly for forest-dependent people.

Keywords: Carbon Credit, Carbon Pricing, Ulumasen Forest

---

[1] The first version of this paper was submitted as part a research essay towards the completion of Master of Forestry at Fenner School of Environment and Society, Australian National university.

Contents

I. Introduction

Emission from deforestation and forest degradation in developing countries estimated to constitute 20 percent (20%) of the total global emission of greenhouse gases (GHG) annually (CIFOR, 2008). Although it contributes a small percentage compare to other sectors that contribute to global GHG emission such as transportation and industries, continual forest depletion in developing countries is considered different given the broad spectrum of other social-economic and environmental impacts it has created and will bring about.

A number of incentive mechanism to reward preservation for environment and services it provides have basically been formulated and in trial, such as Payment of Environmental Services (PES) and Clean Development Mechanism (CDM) under Kyoto Protocol. Recently, Reducing Emission from Deforestation and forest Degradation (REDD) mechanism was proposed and this is basically an advanced version of CDM aimed specifically at curbing deforestation and degradation of forests particularly in developing countries. It basically rewards every tons of carbon preserved in a particular forested areas as a result of preservation activities. The carbon trading deal under REDD mechanism can use non-voluntary mechanism as in those developed under Kyoto Protocol (state-controlled) or voluntary mechanism (free-trade mechanism).

In Indonesia, REDD is viewed as an effective economic incentive for forest conservation efforts and/or against deforestation and forest degradation since most cases of deforestation in Indonesia have been economically driven. In practical, however, this is certainly not straightforward and can be troublesome particularly in countries where deforestation is reflection poor natural resources governances which are underpinned by complex social-political factors and corruption.

This paper seeks to re-inform the government of some of the fundamental constraints that need to be taken into account constantly if REDD is to be implemented effectively and otherwise will create nothing but disincentive for the sustainability of forest management in Indonesia. It first discusses the issue of the leakage, especially how forest and land-use policies may contribute to the leakage issue, which includes natural occurrences such as forest fires. It then looks at potential misuse of funds and its potential in creating a sense to deforest which then extends up to the issues of determining the right beneficiaries and distribution mechanism of REDD fund. Subsequently, the paper continues to discuss the issue of land tenure security and carbon credit pricing. It then ends up with summaries of thoughts and recommendation. This paper argues that the REDD, like other environmental preservation-based credit scheme, is not a panache for forest degradation problems in Indonesia and the government needs to deal seriously with other fundamental causes of forest degradation before REDD can truly be an incentive for forest protection.

II. Methodology

2.1. The Research Methodology: the Thematic Analysis

Methodological approach used in this paper is thematic analysis. Thematic analysis is a method for identifying, analysing and reporting themes of patterns within data (Boyatzis, 1998, Joffe and Yardle, 2004, Braun and Clarke, 2006). Unlike its twin, the content analysis, that tends to describe numerically features of particular texts or images, thematic analysis explores more on qualitative aspects of material analysed (Joffe and Yardle, 2004). Thematic analysis used to be applied more in psychological and clinical trial researches but is increasingly used in others policy research topics (for example, see Agar, 1983).

2.2. Data Collection Method and Analysis

The data collection method used for this paper is mainly document study. Documents reviewed were secondary data and included scholarly publication such as peer-reviewed journals and articles, book chapters and the official government regulations and reports. The analysis was conducted manually, and involves reading the documents, taking notes and assessing them against official reports and government documents.

III. Revisiting Issues of REDD Implementation

3.1. The 'Leakage' Issue

Leakage is one of the fears frequently raised with respect to avoided deforestation. Leakage basically refers to "carbon benefits from a mitigation project being wholly or partially nullified by emission provoked by the project outside of its boundaries (including not only spatial boundaries, but also boundaries in term of time and in terms of conceptual framework) (Fearnside, 2008). The term 'leakage' used in the context of REDD, however, is argued as inappropriate as its definition is often made simplistic and only implies small-scale impact (Globel Forest Coalition, 2008). Nevertheless, it is sensible to say that any methodologies used to calculate green house gasses from reducing deforestation should take into account all forms of leakage both directly and indirectly. This means that such methodologies need to account underlying causes of deforestation and effects (counter-productive) that any forest and land use policies can potentially create themselves. Without doing so, any REDD policies and incentive schemes targeting particular forest areas will unavoidably lead to leakage. To do so, however, can be very difficult, particularly where deforestation is a matter of socio-economic and political complexities and overarching forest and land use policies often contribute to the deforestation processes themselves. The following are some of the leakage issues in Indonesia.

3.1.1. The macro-effect of forest and land-use policies

The macro-effects of forest policies can often trigger deforestation in other areas on a larger scale than that avoided in the specific protected areas. In Indonesia, this has been the case in the effort to reduce pressure over natural forests from illegal logging, which is one the most significant and consistent contributor to deforestation in the country (Sunderlin and Resosudarmo,1996; Obidzinski et al., 2007). *Per se*, illegal logging is primarily caused by inability of production forests to sustainably supply excessive timber and pulps demand and excessive installed capacities of existing wood-based processing industries, in addition to wide-spread corruption and weak law enforcement. The gap between timber demand and supply from 1980-2005 is illustrated in figure 2.1.

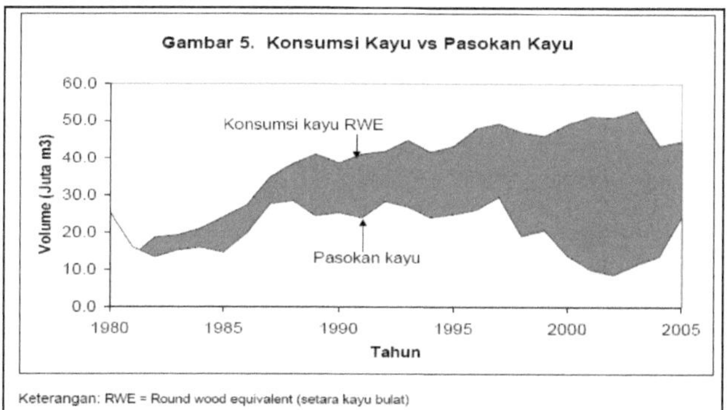

Translation:

Konsumsi Kayu	= Timber Consumption	Keterangan	= Note
Pasokan Kayu	= Timber Demand	Juta	= Million
Juta	= Million	Tahun	= Year

Figure 2.1. Timber Consumption and Demand in Indonesia over the Period of 1980 - 2005

In order to deal minimize the gap and hopefully minimize pressure over natural forests from illegal logging, from 1985-1997 the government put in place log export ban (LEB) policy (which then reinforced in 2003). In its early phase of implementation, this policy had been proven to be very efficient as the frequency of logging operation decreased. Later on, it was realized that the LEB severely impacted on domestic logs price, decrease the value of tree stands in forest and subsequently encourage conversion of forest to more commercial non-forest use (Manurung, 2004). This is also precisely the case in the recent issue of trans-border illegal logging between Malaysia and Indonesia. Although the Government of Malaysia had put in place logs and sawn timber import ban from Indonesia, and the Government of Indonesia increased military operation for law enforcement along the border, the rate of illegal logging is still high (Obidzinski et al., 2007). This proves that policies to halt logging in one area often inevitably promote logging in other areas, unless overall timber consumption and logging and pulp processing capacities are diminished.

3.1.2. Commercialization of agro-industrial monoculture products

The expansion of large-scale, agro-industrial monocultures for food, fiber and energy production is an important cause of deforestation, both directly and indirectly. For example, Indonesia is currently the largest palm oil producer with 6.78 million hectares of plantation (12.3% annual growth). Policy incentive and economic incentive (market and profit gain) offered in this venture have underpinned the rapid growth of palm plantation in Indonesia (both in public and private domain). Figure 2.2 illustrate the development of palm-oil plantation in Indonesia from 1980-2007. Enabling incentive for palm-oil plantation development has been Indonesian Government's Policy that targets the establishment of 6.5 million hectare of sawit plantation throughout the country (mainly in Kalimantan, but now spread to Sulawesi and Papua), and provide conducive licensing mechanism for the development of plantation and processing plants (McCarthy, 2008). Economic incentive of attractive CPO price ($570/m3) encourages farmers to chop down tree stands on their farmland to be replaced by sawit (McCarthy, 2008). It is believed that there have been a large extent of productive home gardens of *Dayak* people (Indigenous Borneo) being converted into sawit plantation (World Rainforest Movement, 1998). Indonesia is obviously intended to maintain its domination in world palm oil market and this means continuos plantation expansion and possibly further forest conversion. To what extent incentive from

REDD can overshadow incentive from agro-industrial monocultures for food, fiber and energy production and thus halt deforestation is too early and difficult to estimate.

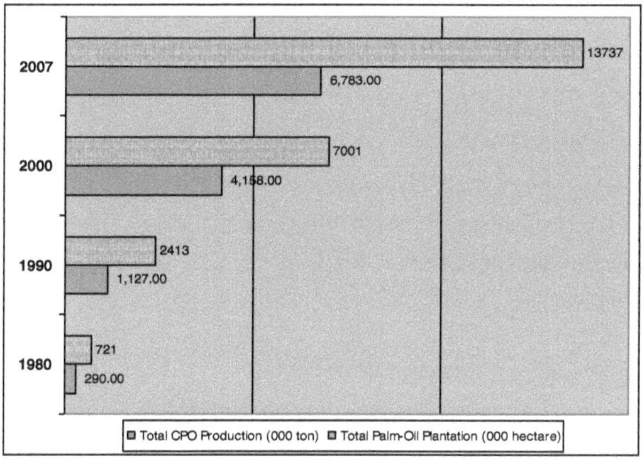

Source: Indonesian Department of Agriculture

Figure 2.2. The Development of Palm-Oil Plantation and Total Production of Crude Palm Oil from 1980 - 2007

3.1.3. Unpredictable Impact of Traditional Use of Fire

Forest fires can lead to significant deforestation, especially in places already degraded by logging or previous fire. There are also concerns about the stability of peat deposits in the region's swampy, lowland forests. Peat—thick layers of partially decayed vegetation—is both extremely flammable and a repository for carbon. Peat burning releases not just carbon monoxide, but also carbon dioxide, a greenhouse gas.

Human activity and natural cycles are likely to conspire to create devastating fire seasons in Indonesia (NASA, 2007). Farmers have been traditionally using fire (Slash and burn method) primarily for agricultural land preparation (as commonly practiced in shifting cultivation tradition). In some other cases, fires are often lit intentionally to maintain the grasslands that local people use to sustain a variety of livelihood activities such as cattle rearing, hunting and farming (Tacconi and Ruhiyat, 2006). Because the use of fires often reduces the use of labor, fire use is often a significant contributor to the local economy (Tacconi and Ruhiyat, 2006). The problem happens when fires escapes into the forest and the farmer lacks of knowledge of fire management. Fisher (1969 cited in Tacconi and Ruhiyat, 2006) thought that farmers did not have ways to stop grassland fires from burning secondary forest.

Future forest fire policies in Indonesia need to take into account human livelihood dimension (social and economic) and the geographic variation in fuels, climate and land use. Otherwise further deforestation perpetuates which means more 'leakage'.

3.1.4. Perverse Right to Deforest and Misuse of PES Funds

Another example of leakage is exemplified in the case of payments of environmental services (PES). The case of PES implementation in Brazil exemplified that there are pervasive issue of 'right to deforest' and/or the misuse of PES fund to actually fund deforestation (Fearnside, 2008). In the recent report released by Center for International Forestry Research Wunder (2008) describes that:

> "...a farm-level payment for environmental services (PES) programme may reward the landowner for not deforesting the PES-enrolled forest plot A during five years. However, if the owner shifted all planned deforestation from plot A to another, non PES-enrolled plot B, mitigation would be entirely offset by leakage or 'displacement of emissions', as the phenomenon is called in the Bali Action Plan (Thirteenth Session of the Conference of the Parties-COP 13). If the landowner further used all PES funds to buy chainsaws to enable additional clearing and cattle to graze on the land, medium-run leakage may well exceed 100 percent of mitigation - implying leakage also has a time dimension, depending on how quickly economic and biophysical processes work. Conversely, if the landowner invested the money in ecotourism or agroforestry and stopped all clearing, leakage would be reversed, crowding in off-site mitigation gains beyond target plot A." (p.66)

This implies that a country may face the risk of losing most of its forests unless it is able to pay every square meter of existing forest. In the case of Leuser National Park, for example, Baukering et al (2006) calculates that environmental services provided would be around US$316 million/year and this is more or less PES incentives that the government have to provide to guarantee that the forest will remain intact until the end of commitment period. Unless if the government can anticipate with ways to ensure that any kind of PES funds, include REDD funds, can be allocated properly potential forest loss due to perverse rights to deforest will still be there. This then lead to next questions as to how to determine the right beneficiaries of REDD funds and what kind of distribution mechanism to be used to ensure efficient and fair allocation of REDD funds.

3.1.5. Determining the Right Beneficiaries

Learning from Payment of Environmental Services (PES) Implementation, concerns have been raised as to whether REDD payment is going to be used for paying the rich landholders instead of used for more socially desirable ends that would reduce the inequalities in wealth distribution (Fearnside, 2008). Or, whether the payment is going to be used to pay actors who have been destroying the forest while traditional forest peoples who have much less impact are left with no reward for the environmental services of the forest (Fearnside, 2008).

In Indonesia, such questions can be very critical in determining who are supposed to receive payments from REDD financing mechanism. For example, the questions who are the actors that destroy the forest are often complex and political. In some cases, logging operators have been blamed for destroying the forests which are the source of livelihood for forest-dependent people. This is the case where forest is actually inhabited by tribal indigenous people. In other cases, however, individuals or groups within communities (rural people) are often involved in forest destruction and this is often the case where people have limited access to alternative income and employment opportunities. While identifying the right beneficiaries is important, just the same important is the question as to whether not to involve embrace these forest destroyer will be more effective then not embracing them? Again there is no panacea for this. However, it will perhaps to start from looking at how strong institutional arrangement and social capital in a particular context.

District and sub-district government involvement should also be taken into account. District and Sub-District Governments are often blamed for deforestation cases. However, it should be noted that district and sub-district governments are government apparatus that stay in the middle access to resource and access to financial. Resource Governance and Fiscal Decentralization in Indonesia are not full decentralization; some of the strategic resources are still centrally controlled. In some cases, lack of power and financial resources to manage the resources have made district and sub-district governments are less empowered. It is therefore important to carefully consider and involves them into the processes. Examples of forest management that is effective due to the strong role local governments exemplifies in the case of Community Forestry in Lampung Barat as well as in the case of Community Teakfarm Management in Southeast Sulawesi (for example Barr, 2006).

3.1.6. Efficient and Fair Payment Distribution Mechanism

For the REDD payment to be effectively allocated to its beneficiaries efficient and fair distribution mechanism is imperative. In its review over 25 "Readiness Plan Idea Notes" (R-PINs) from the World Bank's Forest Carbon Partnership Facility, World Resource Institute notes that "Developing a system for managing REDD revenues will be a significant endeavor for most countries, which is largely unaddressed in the R-PINs." (REDD-Monitor, 2009).

REDD payment distribution depends on the scale of REDD projects. Angelsen et al. (2008) divides REDD scales into three options (as illustrated in Figure 3.1)

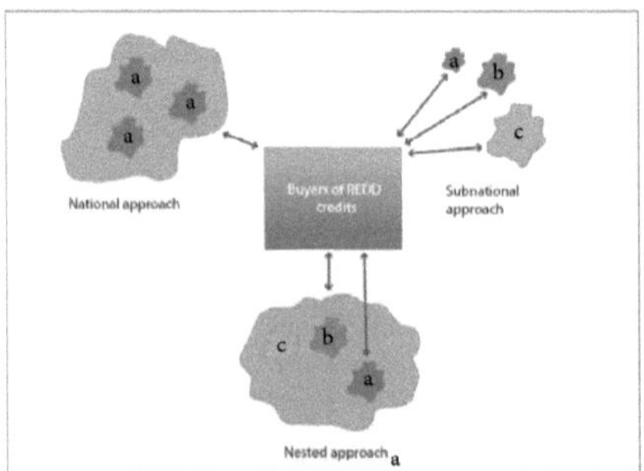

Adapted from: Angelsen et al. (2008)

Figure 3.1. The three approaches to REDD accounting and crediting (note arrows indicate money from the international buyers, and information from (sub) national entities)

Indonesian has had sufficient experiences in managing fund and market incentive (see appendix 1). From these experiences there are a number of factors which often impede efficient distribution mechanism. In a report draft made by Indonesian Forest Climate Alliance, Purnomo et al., (2007) summarizes that these impediments can be one of or combination of the following: 1) Lack of transparency and accountability and high corruption; 2) Opportunities for rent seeking and high transaction costs related to long transaction chains; 3) Perverse incentives; 4) Elite capture and in-migration into areas where incentives are offered; and 5) Lack of demand. Purnomo et al. (2007) then suggests that:

“the more levels in the distribution system the more likely there will be opportunities for rent-seeking and increased inefficiency”. It may therefore be favorable for REDD transactions to occur as close as possible to stakeholders implementing or involved in policies and measures to reduce deforestation and degradation.

If this suggestion is to be followed, than sub-national approach would be best fit Indonesian context. However, it should also be noted that even in REDD sub-national approach (see Figure 3.2, option 2), there are still three possible routes in which fund can be redistributed, namely from donor-provincial government, donor-district government or donor-village/companies. Each of this option will be fit in to particular context. For example, where community institutions are strong and well-established, level of knowledge of the communities is adequate, donor-village route might be chosen; in this case, the government at district and provincial level may act as facilitator which provides technical and if possibly financial support to capacity building of the communities. Otherwise, the other two alternatives may be opted.

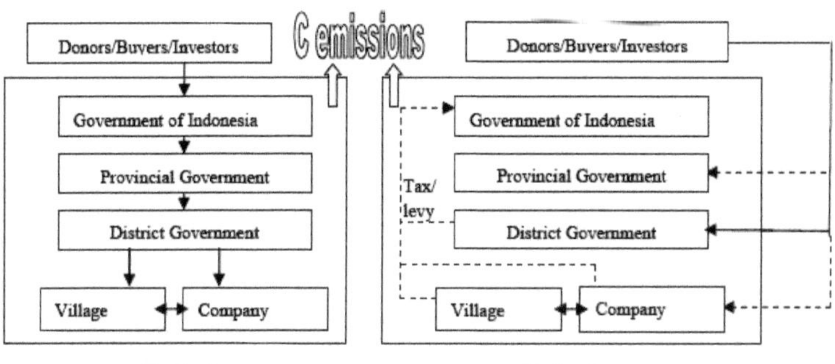

Adapted from Purnomo et al. (2007)

Figure 3.2. Possible REDD mechanims and configurations with buyers/donors/investor

3.2. Land Tenure Insecurity Issue

Overlapping land tenure arrangements between state laws and traditional norms have weakened law enforcement in forest management in Indonesia (Obidzinski et al., 2007). This, in most cases, creates perception of ‘open access’ regarding forests and forest resources which might explain why almost all stakeholders at local and national level are involved in.

Minor recognition over traditional right, less representation of traditional communities in forest management and lack of capacity have contributed to insecure property rights. This is often exacerbated by inconsistent and overlapping land use policy. For example, Presidential Decree No.36/2005 regarding 'Land Procurement for Development Implementation of Public Interest', despite of on-going debate upon it, has been used as a reason for development to take over communities land and reallocate them for another purpose such as for infrastructure development (Asgart, 2005).

Farmers' decision to maintain or convert forest areas highly depends on their perception of security. This was the case of *Damar A*groforestry in Sumatera (see Kusters et al., 2007). It is important that REDD policies deal effectively with the problem of insecure property rights if long term maintenance of forest is to be ensured.

3.3. Carbon Credit Pricing Issue

Dhakal (2009:68) notes that "The carbon market under REDD will be quite different from other market. Payments based on additional units of carbon sequestered are not applicable – that is, no requirement to prove environmental enhancements. Instead, REDD carbon markets sales will provide forest owners with allowances for not using forests in a way that contributes to carbon payments. Payments will be fixed based on references to some economic returns. If this is the case then REDD should ideally be conducted separately from CDM projects or post 2012. Otherwise, both schemes will overlap and which might consequently mean a risk of low carbon price. If the carbon price drops, the question is whether allowance from REDD would be economically competitive than returns form alternative uses of forest and land. Stern (2006 cited in Hall, 2008) calculated that deforestation could be completely eliminated at an average prive of $30 per ton CO2. Seeberg-Elverdelt et al., (2008) suggests that in the case of Payment of Environmental Services in Sulawesi, Indonesia, carbon price of 32€ /ton Carbon will be economically attractive for farmers to adopt a sustainable agroforestry system. Recently, the Californian Government offered the Aceh Province Government a REDD fund of US$9 Million for 4 years commitment (Http://www.redd-monitor.org/2008/11/us-on-the-slippery-slope-to-redd-offsets/), to lock up 750,000 ha of Ulumasen estimated to store approximately 188 ton carbon per hectare (Flora and Fauna International and Carbon Conservation Pty.Ltd, 2007). This amount of fund will provide a carbon price of US$ 0.06/ton or economic return of

US$12/hectare, which is lesser than returns from even a sustainable logging production, which estimated between US$20 – 27 / hectare (see appendix 1). In this case, it is less likely that forest conservation will be benefitted from REDD projects.

IV. CONCLUSION AND RECOMMENDATION

Reducing Emission from Deforestation and forest Degradation (REDD) is one alternative financing mechanism designed to offset forest degradation in developing countries. Since forest degradation in many developing countries is mainly driven by economic motives, REDD can then be proposed as an economic incentive for forest protection. The success of REDD implementation, however, depends highly on whether or not fundamental issues that underline deforestation and forest degradation are solved.

This paper had revisited some fundamental issues it considers as major constraints faced in the efforts to realize sustainable forest management in Indonesia. These include: 1) the leakage issue, which covers aspects such as the macro-effect of forest and land-use policies, commercialization of agro-industrial monoculture products, unpredictable impact of traditional use of fire, perverse right to deforest and misuse of PES funds, inability to determine the right beneficiaries and also efficient and fair payment distribution mechanism, 2) land tenure insecurity issue, and 3) credit pricing issue.

There are potentially a number of other issues whish are not covered in this paper but of similarly importance for example carbon accounting issue. The government also needs to make sure that the baselines for carbon stock calculation is ready before carry on any REDD commitments. This is surely going to be time consuming and not easy given that Indonesia has fragmented ecosystems and each ecosystem definitely has different carbon storage capacity. There is no other way for the government but to increase its investment on research and development.

Eventually, REDD, like any other credit mechanisms, requires creditor to conduct its own feasibility study and fulfill all obligations agreed. If the creditor is confident that the credit will be benefitting, the liabilities can be performed and risks and uncertainties can be managed properly, than it must take the full advantage of such an opportunity. Otherwise, then it perhaps will be better of finding other mechanisms which might pose low risks, particularly to communities that depending highly on the existence of forest resources.

While, it is not yet clear, which mechanism of REDD implementation will be the most effective in terms of distributing benefit of REDD projects, it is recommended that government in provincial and district, as well as facilitator of REDD projects should undertake feasibility study which involve communities of project subsidiaries and announce transparently potential pitfalls if the project is to be carried on. Otherwise, the project will be nothing but further marginalization of local communities, particularly those who have been long dependent on the existence of resources from the forests proposed in the project.

Reference

Agar, M.H., 1983. Political Talk: Thematic Analysis of a Politic Argument, Review of Policy Research, 2(4):601-614

Angelsen (Ed, 2008. Moving Ahead with REDD: Issues, Options and Implications. Center for International Forestry Research (CIFOR), Bogor, Indonesia.

Angelsen, A., and Atmdja, S., 2008. Chapter 1, in Moving Ahead with REDD: Issues, Options and Implications. (Ed). Arild Angelsen, pp 1-9. Center for International Forestry Research (CIFOR), Bogor, Indonesia.

Angelsen, A., Streck, C., Peskett, L., Brown, J., and Luttrell, C., 2008 . What is the Right Scale for REDD, Chapter 4, in Moving Ahead with REDD: Issues, Options and Implications. (Ed). Arild Angelsen, pp 1-9. Center for International Forestry Research (CIFOR), Bogor, Indonesia.

Asgart, S.M., 2005. Dari pembangunan ke penghargaan hak asasi manusia (From development to human rights recognition). Center for Democration and human rights. Source available from:http://www.demosindonesia.org/pdf/Sasgart_dari%20pembangunanisme%20ke%20penghormatan%20HAM.pdf

Boyatzis, R.E. 1998: Transforming qualitative information: thematic analysis and code development. Sage.

Braun, Virginia and Clarke, Victoria(2006) 'Using thematic analysis in psychology', Qualitative Research in Psychology, 3(2): 77 — 101

BAPPEDA (Regional Planning and Developing Agency) Aceh, 2008. Prioritas Pembangunan Ekonomi Aceh (Aceh Economic Development Priority). Presentation of Aceh Province Government to Multi-Donor Forum. Source available froum: http://www.multidonorfund.org/documents/2008-11-03-EDFF_Presentation%20of%20Bappeda%20NAD-.pdf

Barr, R., 2006. Linking Small-Scale Agroforests to International Markets for FSC-Certified Wood: A Case Study of the Cooperative for Sustainable Successful Forests, or 'KHJL'. Source available from: http://www.tropicalforesttrust.com/media/uploaded/KHJL_Case_Study.pdf

Beukering, P.J.H.V., Herman Cesar, H.S.J., and. Janssen, M.A., 2003. Economic valuation of the Leuser National Park on Sumatra, Indonesia. Ecological Economics, 44:43-62.

Dhakal, B., 2009. Carbon Liability, Market Price Risk and Social Impacts of Reducing Emission from Deforestation and Forest Degradation (REDD) Programme. Journal of Forest and Livelihoos, 8(1):67-76.

Hall, R. 2008. Can Forest Carbon Finance Stop Deforestation? A critical review of proposed REDD mechanisms. Friends of the Earth International, Oxford Source available from: www.globalforestcoalition.org/img/userpics/File/publications/Can-forest-carbon-finance-stop-deforestation.pdf

Joffe, H and Yardley, L., 2004. Chapter 4. Content and Thematic Analysis. In Research Methods for Clinical and Health Physicology, P.56-58

Kusters, K., De Foresta, H., Ekadinata, A., and van Noordwijk, M., 2007. Towards Solutions for State vs. Local Community Conflicts Over Forestland: The Impact of Formal Recognition of User Rights in Krui, Sumatra, Indonesia. Human Ecology, 35:4427–438.

Fearnside, P.M., 2008. Amazon Forest Maintenance as a Source of Environmental Service, Anais da Academica Brasiliera de Ciencas, 80(1):101-114.

Flora and Fauna Internation (FFI) and Carbon Conservation Ltd., 2007. Reducing Carbon Emissions from Deforestation in the Ulu Masen Ecosystem, Aceh, Indonesia., Source available from: http://www.climate-standards.org/projects/files/Ulu_Masen_CCBA_Project_Design_Nov1.pdf

Global Forest Coalition, 2008. Potential Policy Approache s and Positive Incentives to Reduce Emissions from Deforestation in Developing Countries, A submission to the Secretariat of the Framework Convention on Climate Change. Source available from: http://www.wrm.org.uy/GFC/material/Incentives_Reduce_Emissions.html

Sundelin, W.D. & Resosudarmo, I.A.P. 1996. Rates and causes of deforestation in Indonesia. Towards a resolution of the ambiguities. CIFOR Occasional Paper No. 9. Bogor. 19pp.

Manurung, E.G.T, 2004. Dampak Kebijakan Larangan Ekspor Kayu Bulat pada Periode 1985-1997 terhadap Sektor Kehutanan Indonesia (The Impact of Log export Ban over The Priod of 1980-1997 to Indonesian Forestry Sector). Source available from: http://www.rimbawan.com/APHI0611/KUMPULAN_TULISAN/2008/Februari/Executive%20Summary-LX-Ban%20-%20final%20-%20TM-%20Des-2002%20.pdf

Manurung, 2007. A Road Map for the Revitalization of Indonesia's Forest Industry. The Forest Industry Revitalization In-house Experts Working Group. Source available from: http://savetheorangutan.org/wp-content/a-road-map-for-the-revitalization-of-indonesis-forest.pdf

Obidzinski, K., Andrianto, A., and Wijaya, C., 2007. Cross-border Timber Trade in Indonesia: Critical or Overstated problem? Forest Governance Lesson from Kalimantan, International Forestry Review, 9(1):526-535.

McCarthy, J. Surfing and Crashing in the Indonesian oil palm bloom. The Jakarta Post, March 30, 2008. Source available from: www.thejakartapost.com/news/2008/03/27/surfing-and-crashing-indonesian-oil-palm-boom.html - 36k

World Rainforest Movement, 1998. Sawit Watch: an Indonesian network against oil palm plantations. Source available from: http://www.wrm.org.uy/bulletin/14/Indonesia2.html

Tacconi, L and Ruhiyat, Y., 2006. Livelihoods, fire and policy in eastern Indonesia. Singapore Journal of Tropical Geography, 27(1): 67-81.

Seeberg-Elverfeldt, C., Schwarze, S., and M. Zeller. 2008. Carbon Finance - Impacts of Environmental Service Payments on Households in Central Sulawesi, Indonesia. STORMA Discussion Paper Series No. 25. May 2008. Georg-August Universität Göttingen, Göttingen

Wunder, S., 2008. How Should We Deal with Leakage?, In Moving Ahead with REDD: Issues, Options and Implications. (Ed). Arild Angelsen, pp 1-9. Center for International Forestry Research (CIFOR), Bogor, Indonesia.

REDD-Monitor, 2009. World Resources Institute review of World Bank-approved R PINs finds critical issues are onspicuously missing_REDD Monitor

Purnomo, H., van Noordwijk, M., Peskett, L., and Setiono, B., 2007. Payment Mechanims, Distribution and Institutional Arrangements (Draft). Indonesian Forest Climate Alliance (IFCA) Study 3: DRAFT prepared for the Ministry of Forestry, Indonesia.

Appendix 1.

Table 1. Simple estimation of carbon price relative to logs value and district government's revenue from logging

Item	Description		Remark
Estimation of the value of logs (wood forest products) from Ulumasen Forest			
A	Size of the Ulumasen Forest	750,000 ha	
B	Assumption of annual increment of Aceh forest	0.53 cu/ha/yr	See Sumarna et al., 2002; also cited in Manurung et al., 2007
C = (A * B)	Estimasted logs potential from Ulumasen forest	397,500 m3/ha	(assumed that forests are managed sustainably, forest' encroachment is minimum and logging impact is minimum)
D	Average market price* of logs of non-commercial species	450,000 IDR	Bagiyono, Pers.Comm., 17 April 2009**
E	Average market price* of logs of non-commercial species	600,000 IDR	Bagiyono, Pers.Comm., 17 April 2009
F = (C*D)	Estimation of logs value if non-commercial species' price is used	178,875,000,000.00 IDR = 17,887,500 US$**	1 US$ = 10,000 IDR
G = (C*E)	Estimation of logs value if commercial species' price is used	238,500,000,000.00 IDR = 23,850,000.00 US$**	1 US$ = 10,000 IDR
Estimation of local government's income from levies and administration fees			
H	Total forest with actual coverage (logs potential) in Aceh	3,200,000 hectare	Aceh Regional Planning and Development Agency (BAPPEDA), 2008
I	Percentage of Ulumasen forest from the total forest area with coverage in Aceh	0.23 %	
J	Aceh Province's annual allocated budget based on regional revenue and proposed budget (APBD) per 2008	940,000,000.00 US$	BAPPEDA Aceh, 2008
K	Contribution of forest sector to Aceh's economy	0.03 %	BAPPEDA Aceh, 2008; Assumed that incomes from non-wood forest products and other non-wood related uses of forests is excluded (since they are often relatively small than income from wood forest products)
L = (J*K)	Contribution of forestry sector to Aceh's economy in IDR	28,200,000.00 US$	
M =	Estimation of provincial government's	10,312,000.00 US$	Assumed that F is used

(L–F)	income from levies and administration costs of forest utilisation in Aceh		
N = (I * M)	Estimation of district government's income from levies and administration fees from forest utilisation in Aceh	2,416,992.19 US$***	Assumption: only wood forest products are counted from Ulumasen forest
O = (F +	**Estimation of the total income of Ulumasen logs value plus district government's income fro levies and administration fees**	**20,304,492.00 US$**	**Assumed that non-commercial log price is used**
P = (O / A)	**Estimated Price of carbon per cubic meter at Ulumasen (if it is to be economically competitive than logging)**	**27.07 US$/Ha**	**Assumed that all transaction costs are internalized into the price**

PS:

* : Exclude CIF price; CIF (cost, insurance and freight)

** : Forest Planner, Forest Research and Development Agency, South Sumatera

*** : This is equal to US$312/hectare, slightly higher than what is suggested by Flora and Fauna International and Carbon Conservation Pty.Ltd (2007:21) which is around $200-2560 per hectare (see http://www.climate-standards.org/projects/files/Ulu_Masen_CCBA_Project_Design_Nov1.pdf